Contents

Ozone and Earth's Atmosphere	4
The UV Index	12
Recipe for Ozone Loss	14
Recipe for Ozone Repair	16
Earth Escape	18
Good Ozone, Bad Ozone	24
Multimedia Information	26
Quick 8 Quiz	28
Learn More	29
Glossary	30
Index	32

A Hole in the Ozone

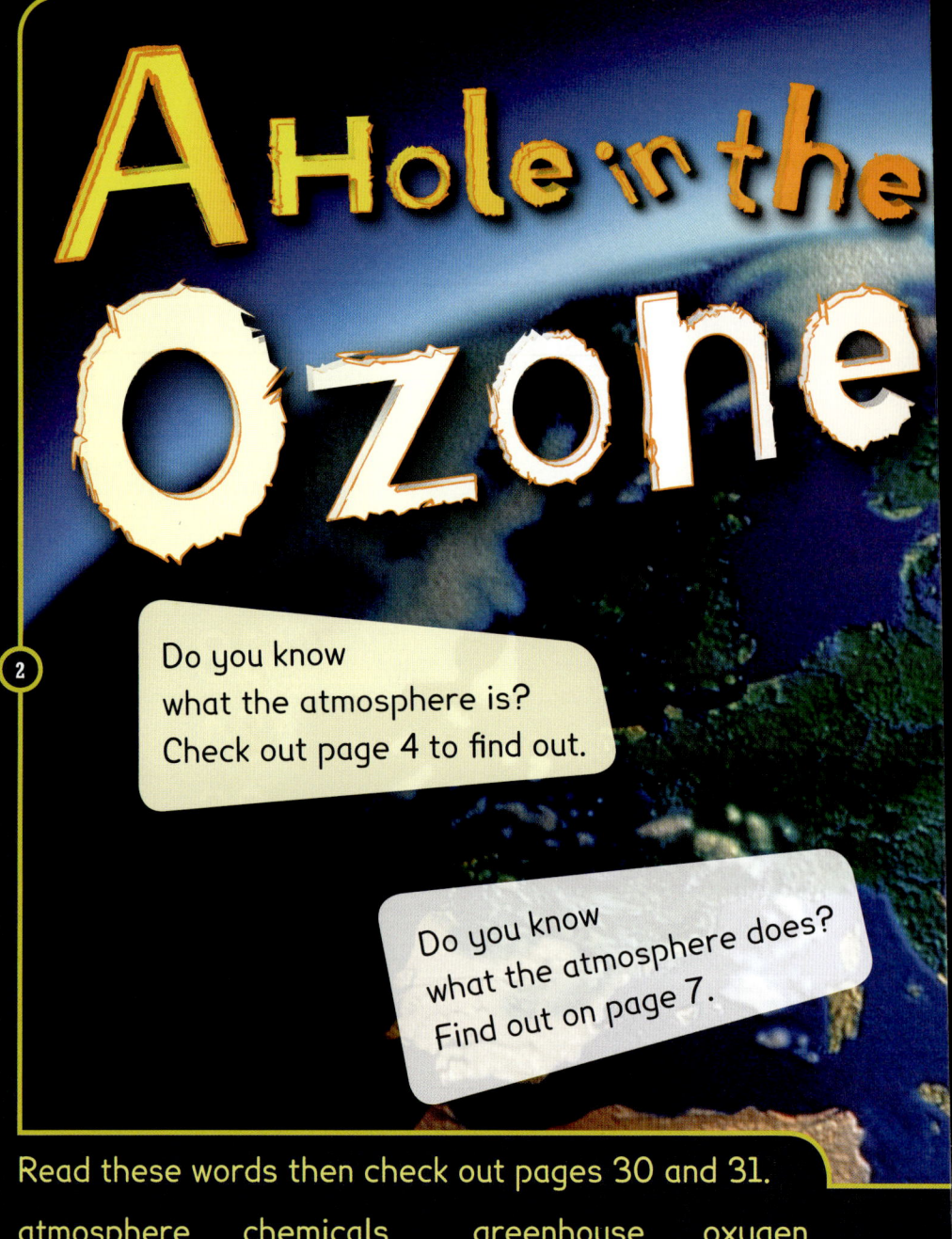

Do you know what the atmosphere is? Check out page 4 to find out.

Do you know what the atmosphere does? Find out on page 7.

Read these words then check out pages 30 and 31.

atmosphere	chemicals	greenhouse gases	oxygen
AT-muhs-feer	KEM-uh-kuhlz	GREEN-hows GAS-uhs	OKS-uh-jen

PURPOSE FOR READING

Did you know that some energy from the sun can hurt you? Find out how on page 8.

Do you know what ozone is? Find out on page 8.

Do you know what makes the hole in the ozone? See page 11.

ozone
OH-zohn

stratosphere
STRAT-uhs-feer

troposphere
TROP-uhs-feer

ultraviolet light
UL-truh-VY-uh-luht LYT

Ozone and Earth's Atmosphere

Written by Collette Manners

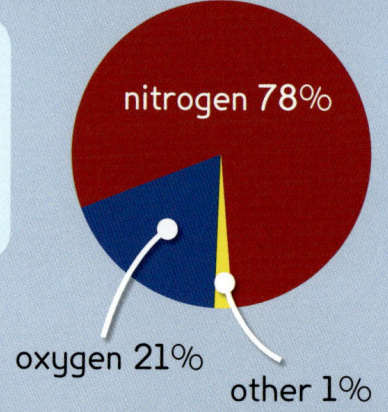

Gases in the Air

nitrogen 78%
oxygen 21%
other 1%

Have you heard people say, "There's a hole in the **ozone**?" Do you know what it means? Do you know what ozone is? To understand, you have to know about the air around you.

Take a deep breath in. Let your breath out. What are you breathing? You are breathing air. Air is made of gases. Air is all around Earth. This air is called the **atmosphere**. It wraps around Earth like a blanket.

This man can see his own breath in the cold air.

But the atmosphere is not only air.
There is water in the atmosphere.
Look up at the clouds.
They hold water.
Wind blows over the land.
It blows dust and smoke into the atmosphere.
Wind blows over the oceans.
It blows water and salt into the atmosphere.

Smoke and steam from Kilauea volcano in Hawaii

There is plenty of air close to Earth.
The air gets thinner higher up.
Then it fades away.
There is no air in space.

The atmosphere is made up of layers.
Look at the diagram.
It shows you the layers.

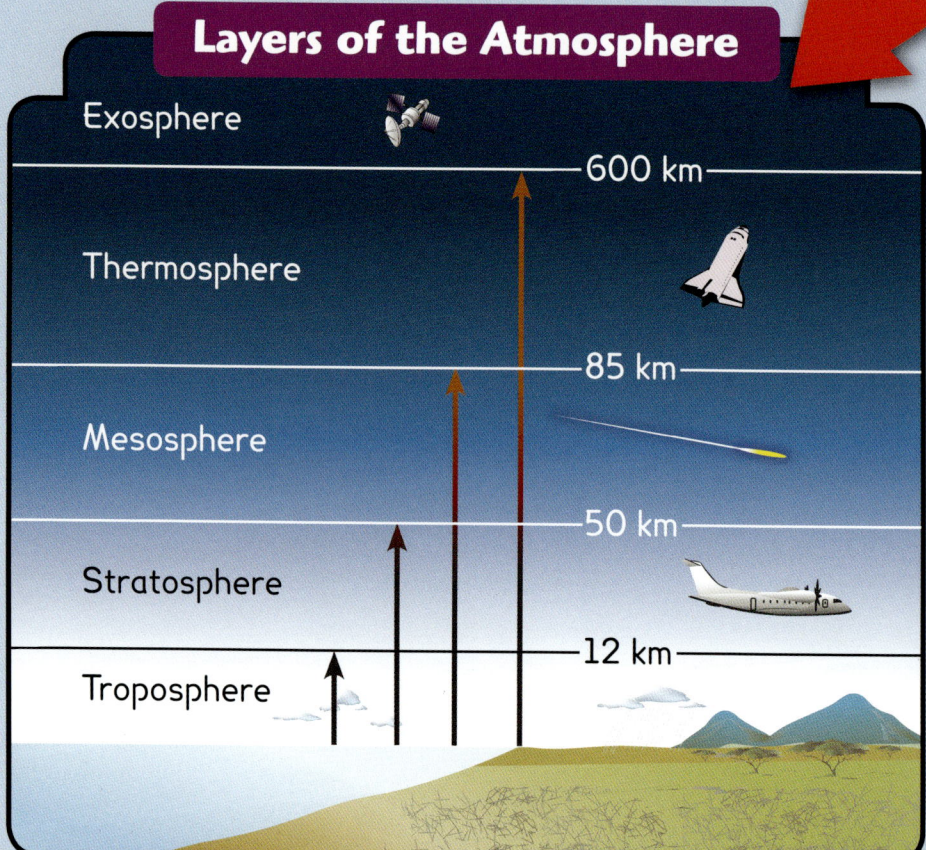

Layers of the Atmosphere

Exosphere
— 600 km —
Thermosphere
— 85 km —
Mesosphere
— 50 km —
Stratosphere
— 12 km —
Troposphere

The atmosphere protects Earth.
It keeps out the cold from space.
It keeps out some of the sun's energy.
The atmosphere traps some of the sun's heat.
This keeps Earth warm, just like a greenhouse.
It is called the greenhouse effect.

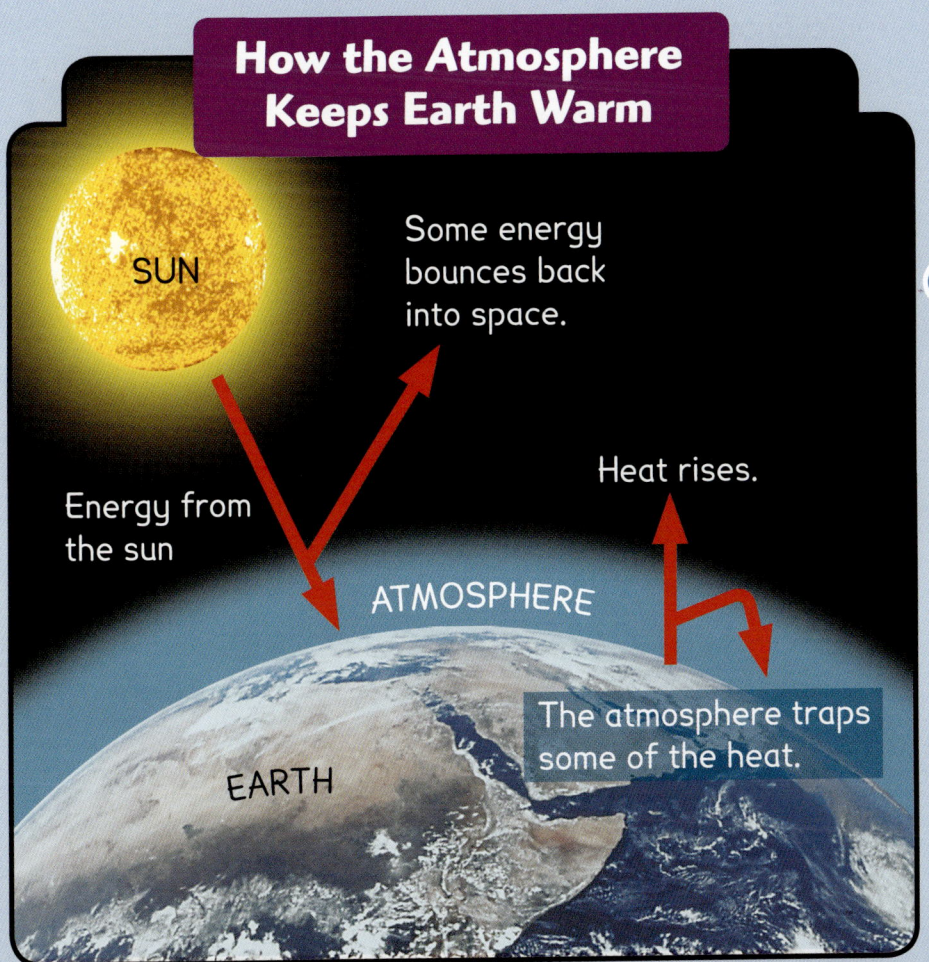

How the Atmosphere Keeps Earth Warm

SUN

Some energy bounces back into space.

Energy from the sun

Heat rises.

ATMOSPHERE

The atmosphere traps some of the heat.

EARTH

Good Energy, Bad Energy

Some of the sun's energy is good.
Plants and animals need the sun's energy to grow.
They need the heat and light from the sun.

Some of the sun's energy is bad.
A kind of light from the sun is bad.
You cannot see this light,
but it can hurt your eyes.
It can give you skin cancer.
It can harm plants.
It is called **ultraviolet light**.

Go to pages 2 and 3 for help to say the tricky words.

Ultraviolet light can be called UV rays.
Most UV rays stay in a layer of the atmosphere
called the **stratosphere**.
They shine on a gas called **oxygen**.
This makes another gas.
And this brings us back to ozone.
The gas made by UV rays and oxygen is ozone.

Ozone gas forms a layer in the stratosphere.
It keeps most of the UV rays away from Earth.

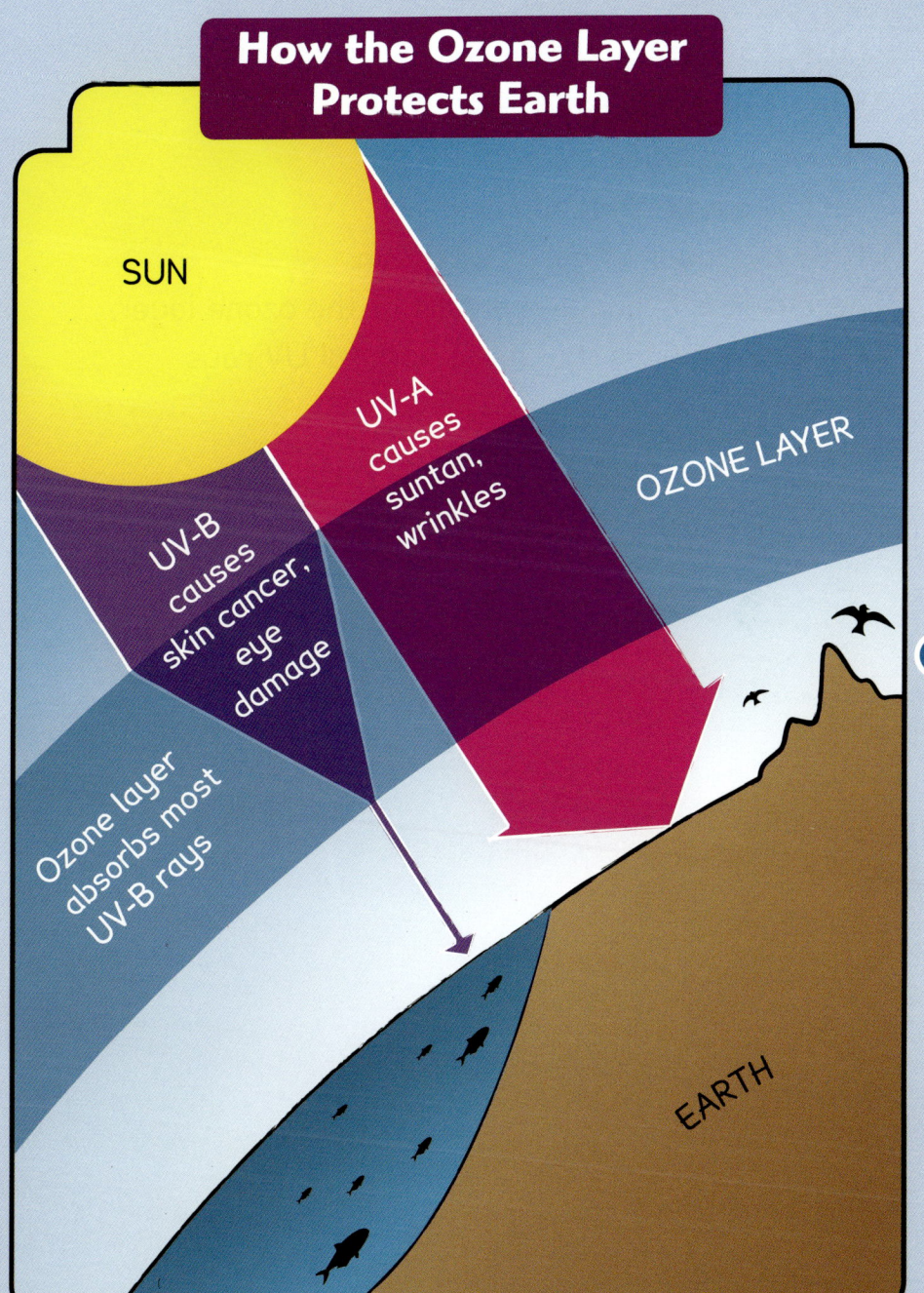

The Ozone Hole

There is never really a hole in the ozone layer.
There is a hole-shaped place where the layer of gas is very thin.
There is less ozone in this part of the ozone layer. Less ozone means that more harmful UV rays can reach Earth.

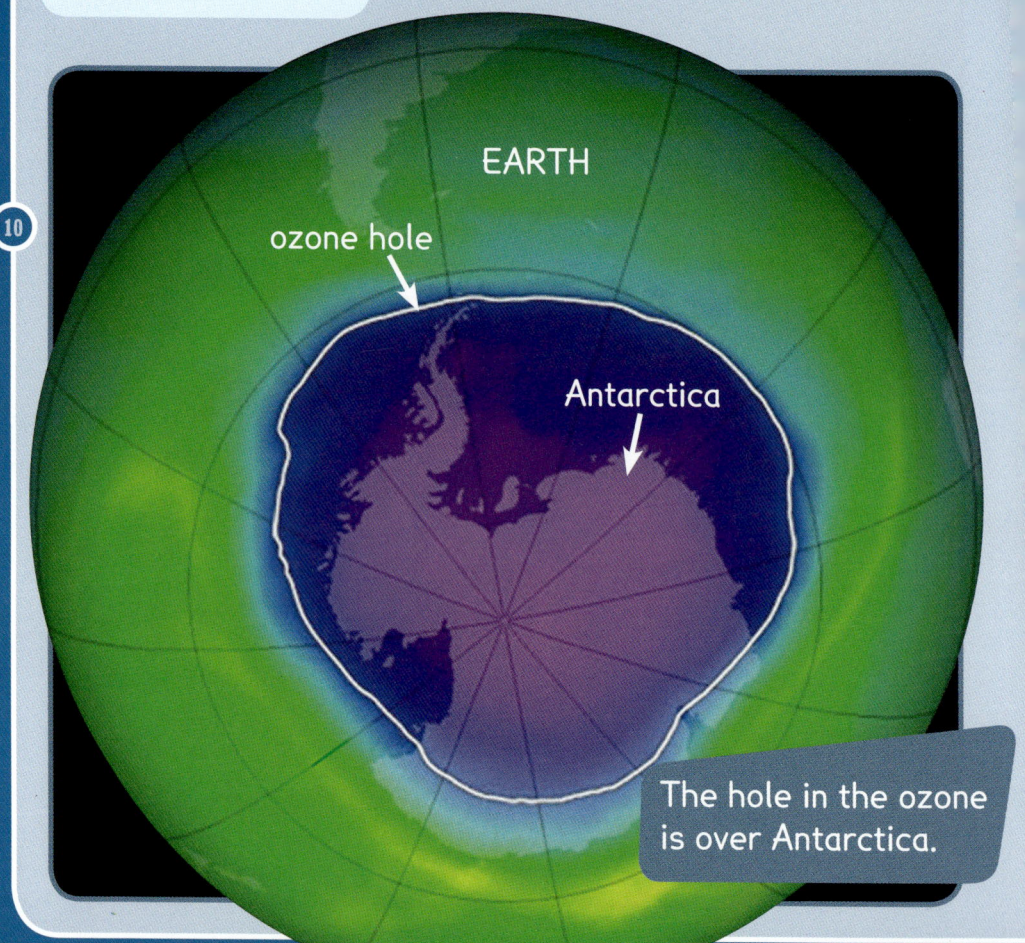

The hole in the ozone is over Antarctica.

Why is there less ozone here?
Because **chemicals** called CFCs
are in the air over Antarctica.
They are in ice clouds that form only in winter.
This is when the sun does not shine here at all.
The CFCs in the clouds are not harmful in the dark.
They need sunlight to start working.

In spring, the sun comes out.
The CFCs start to break down the ozone.
This makes the ozone layer very thin.
The ice clouds go away as it gets warmer.
The ozone layer starts to recover.

Ice clouds form in the stratosphere over Antarctica. Ozone loss happens very fast in clouds like these.

Read on to find out how the UV Index helps people keep safe. →

The UV Index

Written by Collette Manners

The UV Index tells you about UV rays. It tells you how long you can stay out in the sun.
It uses numbers from 1 to 11.

Remember water, sand, and snow reflect UV rays. Wear sunglasses. Use sunscreen.

The hole in the ozone means more UV rays reach Earth.

UV Index
1 2 3 4 5 6 7 8 9 10 11

0–2 Low danger
Wear a hat.

3–5 Moderate danger
Wear a hat.
Use sunscreen.

1 hour — Burn time: more than one hour

6-7 High danger
Wear a hat.
Use sunscreen.
Cover up when you go outside.
Stay in the shade in the middle of the day.

 Burn time: about 30 minutes

8-10 Very high danger
Wear a hat.
Use sunscreen.
Cover up.
Wear sunglasses.
Try to stay in the shade between 10 a.m. and 4 p.m.

 Burn time: about 20 minutes

11+ Extreme danger
Wear a hat.
Use sunscreen.
Cover up.
Wear sunglasses.
Stay inside if you can.

 Burn time: less than 15 minutes

Read on for a recipe for ozone loss. →

Recipe for Ozone Loss

Written by Collette Manners

For this recipe you will need:
People
Spray cans
Fridges
Fire extinguishers

CFCs are chlorofluorocarbons (KLAW-roh-FLOO-roh-KAHR-buhnz).

What to do:

> Start any time after 1930

You have about 60 years to make this recipe.
You will not be able to buy things to make it after that.
But this recipe keeps on working after you stop making it!

Use spray cans.
Spray cans use CFCs
to make what is inside spray out.
CFCs are bad for the ozone layer.
They break down the ozone.

Use fridges.
Fridges use CFCs to keep things cool.
CFCs can leak into the air.
They can stay there for years.

Use fire extinguishers.
Fire extinguishers use CFCs
to pump foam onto fires.
The CFCs go into the stratosphere.
They break down the ozone there.

Mix all these together.
You have ozone loss.

CFCs can stay in the atmosphere for 20–100 years!

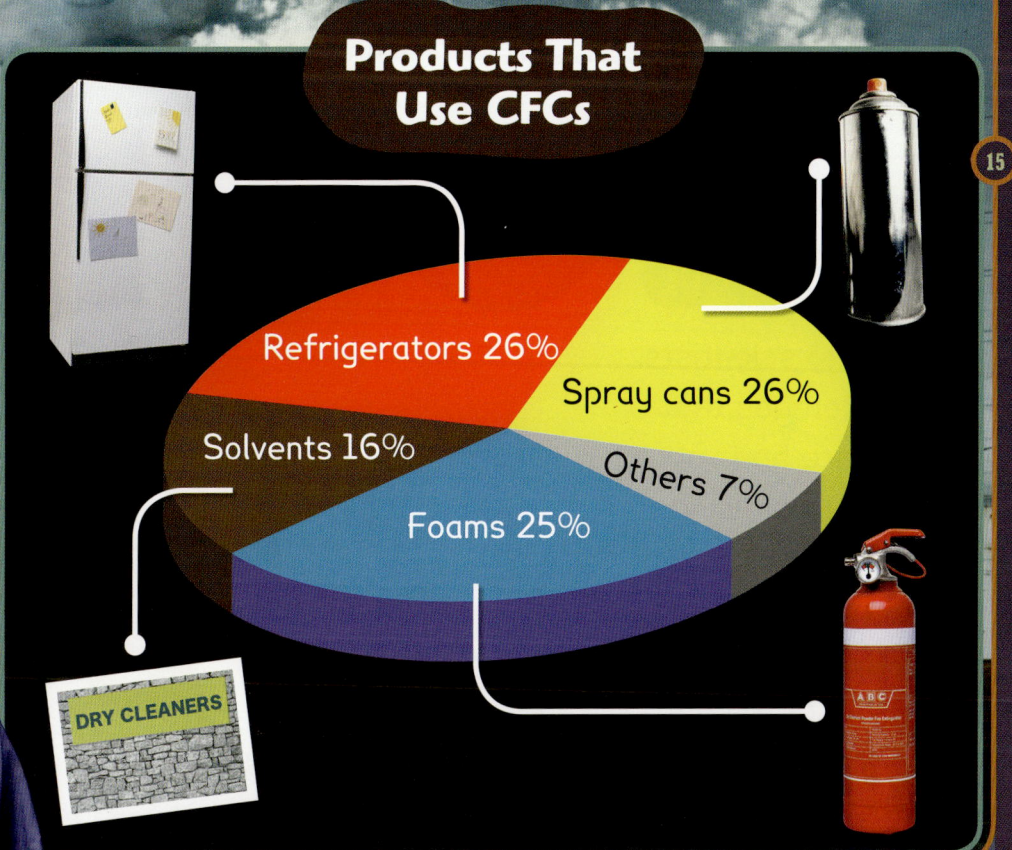

Products That Use CFCs

- Refrigerators 26%
- Spray cans 26%
- Solvents 16%
- Foams 25%
- Others 7%
- DRY CLEANERS

Read on for a recipe for ozone repair. →

Recipe for Ozone Repair

WRITTEN BY COLLETTE MANNERS

For this recipe you will need:
People around the world – include scientists and governments
Information about CFCs and how they thin ozone
New chemicals that are safe for the ozone layer

One CFC molecule can break down 100,000 molecules of ozone.

ozone hole

What to do:

Start in the 1970s

Listen to scientists. They tell you that CFCs are bad. CFCs break down the ozone layer.

Get a shock in 1985

Scientists tell you there is a hole in the ozone.

RECIPE

Meet in 1987

Meet with many countries.
Share what you know.
Make some rules:
1. Stop making CFCs.
2. Use safe chemicals instead.
Agree to keep these rules.

Mix all these together.
Don't stop yet.
You need to keep mixing till about 2070.
You will repair the hole in the ozone.

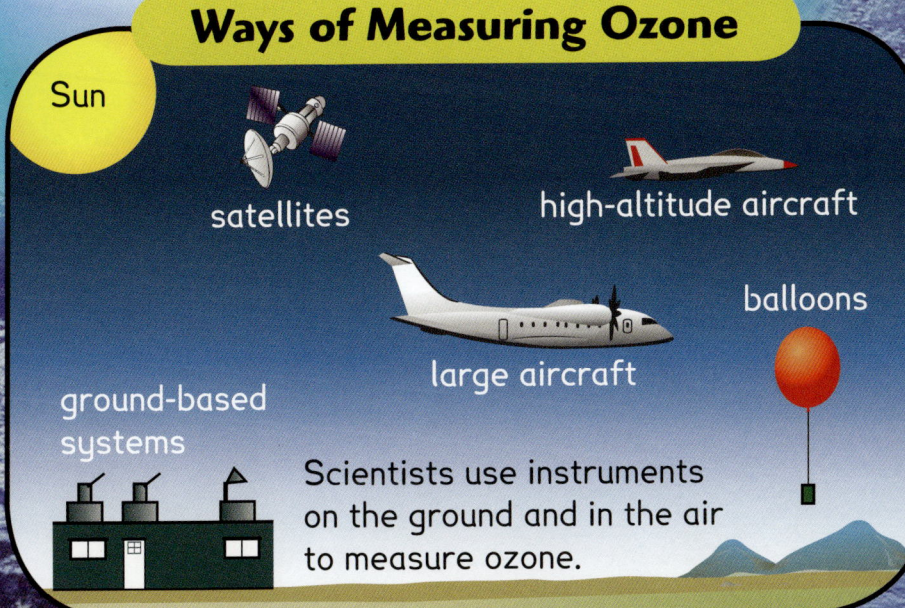

Ways of Measuring Ozone

Sun

satellites

high-altitude aircraft

large aircraft

balloons

ground-based systems

Scientists use instruments on the ground and in the air to measure ozone.

Turn the page for a story about escaping from Earth.

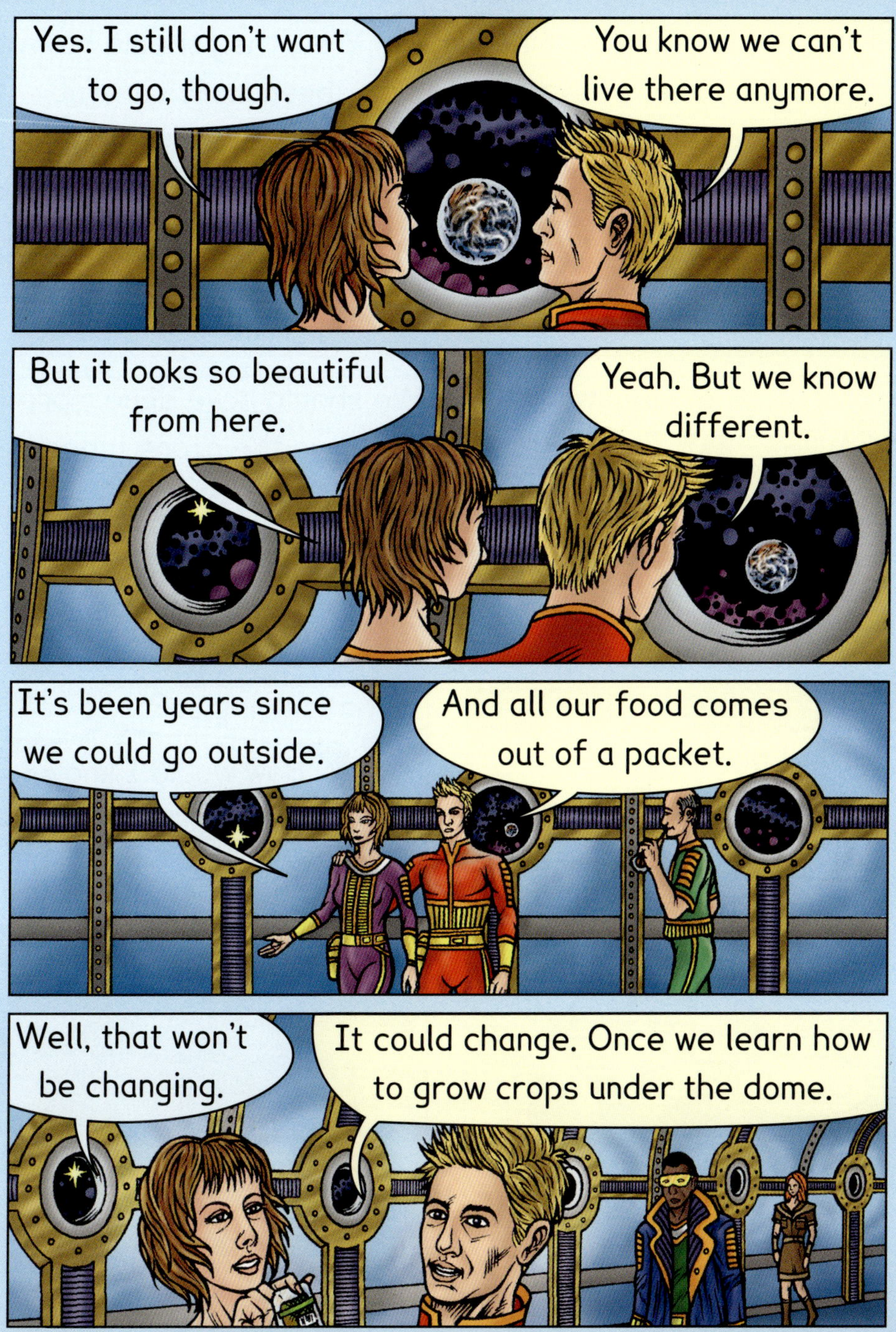

Good Ozone, Bad Ozone

Written by Collette Manners

Do you know that there is good ozone and bad ozone? Good ozone keeps Earth safe from the sun's harmful rays. Good ozone is in the stratosphere.

Bad ozone is closer to Earth. Bad ozone is in the **troposphere**. People make bad ozone. Bad ozone is in smog. It poisons, or pollutes, Earth's atmosphere. Bad ozone is often over cities. But the wind blows bad ozone. It can blow it far away from the city.

Smog over Shanghai, China

Bad ozone makes it hard to breathe.

People can try to stop making bad ozone.
They can burn less petrol and diesel.
They can burn less coal and oil.

Burning fossil fuels such as petrol and diesel makes bad ozone.

What can you do?
You can learn more about bad ozone.
You can work out ways
to help keep Earth clean and safe.
You can tell people what you know.

Bad ozone makes it hard for plants to grow.

INFORMATION PAMPHLET

Read on to find out more about bad ozone.

Multimedia Information

www.readingwinners.com.au

FAQS

Q Can the bad ozone that people make fix the hole in the ozone?

A No.
Bad ozone stays close to Earth.
It is in the troposphere.
It does not float up to the stratosphere.
Bad ozone cannot fix the hole in the ozone.

Bad ozone traps heat from the sun.
Other gases in the atmosphere do this, too.
They are all **greenhouse gases**.
Click here to find out more about them.

Carbon dioxide, methane, nitrous oxide, and ozone are all greenhouse gases.

www.readingwinners.com.au

Greenhouse Gases | **Ozone** | **Atmosphere** | **UV Light**

Greenhouse Gases

Greenhouse gases keep Earth warm so people, plants, and animals can live. This is called the greenhouse effect.

Look back at page 7.

Greenhouse gases are part of nature. But people make these gases, too. They make too many greenhouse gases. Earth gets too warm.

This is global warming.

Let's make Earth safe for living things. Let's make fewer greenhouse gases.

Ice is melting in the Arctic because of global warming.

Turn the page to check what you have learned. →

Quick 8 Quiz

1. Name three things in the atmosphere.
2. Where is the ozone layer?
3. What two things make ozone in the ozone layer?
4. What does the ozone layer protect us from?
5. Where is the hole in the ozone?
6. Name some chemicals that are bad for the ozone layer.
7. Where is bad ozone?
8. What does bad ozone do?

Turn to page 32 for clues. →

Learn More

Choose Your Topic
You want to help keep Earth clean and safe. Choose a greenhouse gas from this book.

Research Your Topic
Find out how people make your greenhouse gas.
Find out ways to stop making it.
Find out ways to make Earth cleaner and safer.

Write Your Article
Make some notes first.
Find pictures.
Find diagrams.
Find maps.
Get your facts in order.
Use subheadings to help you.
Write a draft.
Check your spelling.
Check your punctuation.

Present Your Topic
Share your work with other members of your group.

atmosphere – a blanket of air around Earth, made of gases, water vapour, and tiny particles

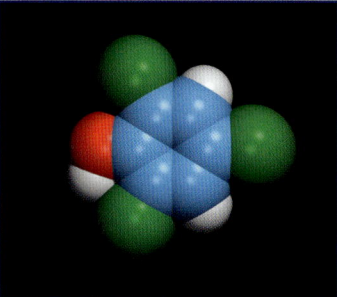

chemicals – gases, liquids, or solids made by people or by nature. Some chemicals are harmful.

greenhouse gases – gases that trap heat from the sun close to Earth's surface

oxygen

oxygen – a gas in air that people need to live and breathe

GLOSSARY

ozone

ozone – a gas that forms when UV rays shine on oxygen

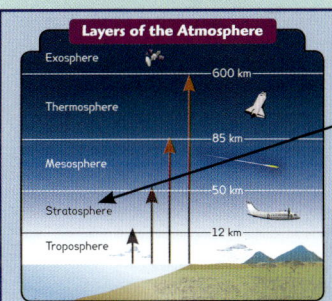

stratosphere – the layer of the atmosphere that is above the clouds

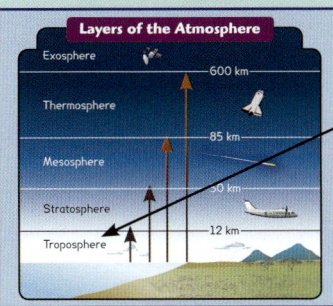

troposphere – the layer of the atmosphere that is closest to Earth

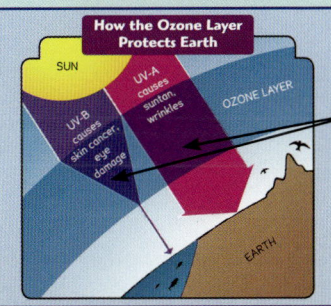

ultraviolet light – a kind of energy that comes from the sun

Index

air 4–6, 11, 14, 17
Antarctica 10, 11
atmosphere 4–8, 15, 24, 26
CFCs 11, 14–17
chemicals 11, 16, 17
clouds 5, 11
Earth 4, 6–10, 12, 18, 20, 22–27
energy 7, 8
gas(es) 4, 8, 10
greenhouse
   effect 7, 27
   gases 26, 27
oxygen 4, 8
ozone 4, 8–11, 14–17, 24–26
   hole 10, 16, 17
   layer 9–11, 14, 16, 20
scientists 16, 17, 22
space 6, 7
stratosphere   6, 8, 11, 15, 24, 26
troposphere 6, 24, 26
ultraviolet light . . . 8
UV Index 12
UV rays 8, 10, 12

Clues to the Quick 8 Quiz

1. Go to page 5.
2. Go to page 8.
3. Go to page 8.
4. Go to page 9.
5. Go to page 10.
6. Go to page 11.
7. Go to page 24.
8. Go to page 24.